Advances in Scientific Knowledge

Miguel A. Sanchez-Rey

Global Policy and the Scientific Age in the Modern Era

Policy Proposal

Version 1

And

Version 2

Global Policy and the Scientific Age in the Modern Era

By

Miguel A. Sanchez-Rey

President Barack Obama's policy decisions is indicative of a failed policy era. In particular, his response to U.S. climate change, trade and welfare spending has not produce the kind of response that much of the world, including in the developing world, has implemented, in order, to achieve a lasting shift that would invigorate the public to achieve an overwhelming vote that would give a unanimous control of the House of Congress to the Democratic Party. The ineffectiveness of the House of Congress has also been ominous. But even if the Democrats were to take control of the House no drastic changes to political consciousness would surmount only that previous policy decisions are to be protected as not incite private interests.

As President Barack Obama has resorted to the executive order to implement popular policy sidestepping the stipulation of Checks and Balances that has been the staple of the Constitution. Unleashing a little, then notice, constitutional crisis in the modern era.

The party Democrats, including the current U.S. presidential candidate Hillary Rodham Clinton, has restated equal policy choices that are as centrists as, "Barack," and yet essential, but ineffective in guiding global policy toward normalized relations of economic necessity or geopolitical concerns.

Markets are motivated by short-term interests. Governments exist to protect the public from the ravishes of the private sector. Current climate policy must, and inevitability if either/or Donald Trump or Hillary Clinton were to take the Executive Office, adhere to private interests if it is to be enacted. And those interests means that short-term profits come first than the welfare of the public in general.

Sustainability is a promising endeavor and catastrophe. One could ever hope that technological advances in solar energy, windmills, electric cars and etc., would, in the long-term, bring an end to rising ocean temperatures and slow-down the pace in which warmer climates has surmounted.

Even then much of the world population depends on a minimal resource of food, products and services. Those minimal resources are ever more shorten by the constant depletion of minerals in which there is, at the time, no resolution to the crisis and contradictory nature of using ITER [International Thermonuclear Experimental Reactor] to achieve a utilitarian society is to be use to control and civilize the global population.

Its ever more obvious, that to use ITER efficiently, requires mass-production. But its downsides are that mass cooperation must ensue but mineral depletion becomes inevitable.

So its resolution, in the Scientific Age, is to implement PHPR [The Physicalist Program] as a resolution to a foreseeable catastrophic scenario in the form of the task. That task is the First Task of PHPR which is how to achieve a terraformic reaction. But to even consider doing so requires a serene world in which the First Task of PHPR is to have a 40-year window to complete 60 percent and another 60 years to fully complete. Those 40 years must also pursue a stand-down of all military hostilities, at a world-wide scale, that leads to a quite and serene state in which a global decline is to be of little notice and in which completion of the First Task of PHPR achieves a vibrant society similar to the 90's era that came before the dawn of Scientific Age.

It makes little difference if the Presidential office is to be held by either a Democrat or Republican. The executive order, that has been widely used to bypass Congressional deadlock, has become the norm of national policy. Either/or the executive order will continue to frame policy decisions that the House of Congress is to continue to be an ineffective and debating body in which deadlock is to continue for the foreseeable future. That said U.S. national defense is independent and rising terrorism will continue to plague world-affairs as long as globalist policies continue.

How to establish ITER is a controversial riddle that issues of genocide become the lime-light. If the world population is to use ITER mineral resources must be ration. If that's the case, to meet rising population growth and rising demand, ITER, essential to controlling climate change, is to be restricted. If restriction is necessary than the population is to be minimally deprived though economic growth is to expand.

Much of the population is then to adhere to Western interests since those interests are known to be more stable and progressive. Violence, declining in the modern era, is to enacted as much of remaining population becomes resistant to powerful private interests as the commons become ever more threaten by rising scarcity and the fruition of a stable democratic order in the Scientific Age.

How to relate PHPR and military interests must also consider national policy. The unstable political process of the modern era renders PHPR top-secret. Where advances in the First Task is to be kept classified until completion is in the near horizon as a scientific dictatorship becomes more evident. A scientific dictatorship that uses the executive order to implement popular policy and in which the House of Congress debates popular policy. Policies that are popular to market forces and cognizant to mineral restrictions. That said the executive office and the House of Congress, which implies the state-capitalist sector, it to be permanently shut-out and in which utmost secrecy must be kept to prevent war-crime and a fall-out due to metaspace.

An international effort must also be of consideration. To pursue an international effort national laboratories must pull their research repositories from public knowledge and a lock-down of all scientific advances relevant to PHPR, and the Scientific Age, is to follow until full-disclosure becomes relevant to the scientific process.

Control of public option, in order to prevent havoc to short-term and long-term strategic interests, is to pacify the world-population through mass economic growth and free education that puts heavy restrictions to scientific participation. A sacrifice in which the pay-offs are limitless and in which its duration is to be minimal.

Policy is an efficient means to meet standards and tranquility. The policies of the modern era is to be open to policy efficiency and productiveness. Mass-production of ITER must, as well, be a productive and efficient endeavor. Depravity is the current reality of U.S. national interests and of concern to intergovernmental establishments. Decline becomes true but, nevertheless, social and natural tranquility is determined to be the norm.

Global Policy and the Scientific Age in the Modern Era

By

Miguel A. Sanchez-Rey

President Barack Obama's policy decisions is indicative of a failed policy era. In particular, his response to U.S. climate change, trade, and welfare spending has not produce the kind of response that much of the world, including in the developing world, has implemented, in order, to achieve a lasting shift that invigorates the public to achieve an overwhelming vote that gives a unanimous control of the House of Congress to the Democratic Party. The ineffectiveness of the House of Congress has also been ominous. But even if the Democrats were to take control of the House no drastic changes to political consciousness surmounts only that previous policy decisions are to be protected so as not to incite private interests.

As President Barack Obama has resorted to the executive order to implement popular policy sidestepping the stipulation of Checks and Balances that has been the staple of the American Constitution. Unleashing a little, then notice, constitutional crisis in the modern era.

The party Democrats, including the current U.S. presidential candidate Hillary Rodham Clinton, has restated equal policy choices that are as centrists as, "Barack," and yet, essential, but ineffective in guiding policy toward normalized relations of economic necessity or geopolitical concerns.

Markets are motivated by short-term interests. Government exist to protect the public from the ravishes of the private sector. Current climate policy must, inevitably, if either/or Donald Trump or Hillary Clinton were to take the Executive Office, comply with private interests if it is to be enacted. And those interest means that short-term profits come first than the welfare of the public in general.

Sustainability is a promising endeavor and catastrophe. One only hopes that technological advances in solar energy, windmills, electric cars, and et., could, in the long-term, bring an end to rising ocean temperatures and slow-down the rate in which warmer climates has surmounted.

Even then much of the world population depends on a minimal resource of food, products, and services. Those minimal resources are shorting by the constant depletion of minerals in which there is, at the time, no resolution to the crisis and contradictory nature of using ITER [International Thermonuclear Experimental Reactor] to achieve a utilitarian society. A utilitarian society that controls and civilizes its own population with clean-energy.

Its ever more obvious, that to use ITER efficiently, requires mass-production. But its drawbacks are that mass cooperation must ensue but mineral depletion becomes inevitable. So its resolution, in the Scientific Age, is to fully implement PHPR [The Physicalist Program] as a resolution to a foreseeable catastrophic scenario in the form of a task. The First Task of PHPR is how to achieve a terraformic reaction. But to even consider doing so requires a serene planet in which the First Task of PHPR is to have a 40-year window to complete 60 percent and another 60 years to reach full completion. Those 40-years must also pursue a stand-down of all military hostilities, at a world-wide scale, that results in a quiet and serene state, in which, a global decline is to be of little notice and in which completion of the First Task achieves a vibrant society similar to 90's era that came before the dawn of the Scientific Age.

It makes little difference if the Presidential office is to be held by either a Democrat or Republican. The executive order, that has been use to bypass Congressional deadlock, has become the norm of national policy. Either/or the executive order will continue to frame policy decisions that the House of Congress is to continue to be an ineffective and debating body in which deadlock is to continue for the foreseeable future. That said U.S. national defense will continue to remain an independent entity and rising terrorism will continue to plague world-affairs as long as globalist policies continue.

How to establish ITER is a controversial riddle that issues of genocide become the lime-light. If the world-population is to use ITER mineral resources must be ration. If that's the case, to meet rising population growth and rising demand, ITER, essential to controlling and mitigating climate change, is to be restricted. If restriction is necessary than the population is to be minimally deprive though economic growth is to expand.

Much of the population is then to obey Western interests since those interests are known to be more stable and progressive. Violence, though declining in the modern era, is to be enacted as much of the remaining population becomes resistant to powerful private interests. Where the commons become ever more threaten by rising scarcity and the fruition of a stable democratic order in the Scientific Age.

How to relate military interests to PHPR requires consideration of national policy. The unstable political process of the modern era renders PHPR top-secret. Where advances in the First Task is to be kept classified until completion is in the near horizon as a scientific dictatorship is evident. A scientific dictatorship that uses the executive order to implement popular policy and in which the House of Congress debates popular policy. Policies that are popular to market forces and cognizant to mineral restrictions. That said the Executive Office and the House of Congress, which also implies that state-capitalist sector, is to be permanently shut-out and in which the utmost secrecy must be kept to prevent a fall-out due to metaspace and war-crime.

An international effort must also be of consideration. To pursue an international effort national laboratories must pull their research repositories from public knowledge and a lock-down of all scientific advances relevant to PHPR, and the Scientific Age, is to follow until full-disclosure becomes relevant to the scientific process.

Control of public option, in order to prevent havoc to short-term and long-term strategic interests, is to pacify the world-population through mass economic growth and free education that puts heavy restrictions to scientific participation. A sacrifice in which the pay-offs are limitless and in which its duration is to be minimal.

Policy is an efficient means to meet standards and tranquility. The policies of the modern era require openness to policy efficiency and productiveness. Mass-production of ITER must, as well, be a productive and efficient endeavor. Depravity is the current reality of U.S. national interests and of concern to the intergovernmental establishment. Decline becomes true but, nevertheless, social and natural tranquility is determined to be the norm.

The Scientific Age: Protocol

By

Miguel A. Sanchez-Rey

The Scientific Age is an uncontrollable scientific machine.

Gaining control of the scientific machine requires that at a certain point the general public be periodically shut-out from the Scientific Age so as not to cause any further harm to the scientific process.

All national laboratories research repositories are to be withheld from the general public. All scientific advances relevant to PHPR, and the Scientific Age, is to follow until full-disclosure becomes relevant to the scientific process.

Knowledge of the Advance Age is to be ascertain and establish by the Grandmaster in anticipation that the last task will be completed.

The Scientific Age leads to the Advance Age when the scientific state is dismantled. At that point the scientific machine becomes controllable.

The Advance Age is an age of wild strength.

Logical Form In Favor of Long Equations

Miguel A. Sanchez-Rey

Abstract

We aim to produce systemic analysis that produces data efficiency. Data efficiency that requires sacrificing over-emphasis of LF [Logical Form] in favor of solutions that resonates in the classical approximation and statistical quantization.

September 12th, 2016

N [variant [of stringy]] is equivalent to perfect number P_N. [1] Homotopic space that is denoted as $H_M \Longrightarrow$ continuous, one-to-one, onto, and infinitely differentiable in finite topology [2]. Such that N [variant [of stringy]] is of perfect number P_N. In which $P_N \subseteq H_M$; where: $\{ \exists W_{ij} \exists T_{kl} \}$ is an element of P_N. We introduce LF as the definition of PHPR [The Physicalist Program] as p [n]$\longrightarrow$ p of essential control [3]. Implying computational control to regulate and parameterize to achieve calculating efficiency in imaginary space. The question is how do we produce systemic analysis? The problem addresses the relational shift that is data analysis due to the quantum measurement problem and the classical approximation that pertains to metaspace. Data analysis that is intrinsic to statistical quantization. We start by ascertaining the holographic counterterms: the Wilson operator and its derivative that merges all five string theories, through mirror symmetry and compactification into 11-dimensional hyperspace, that is as the first known variant [of stringy], as the Super-Yang Mills Gauge Analog [4]. Suppose we have that Lagrangian $[\mathcal{L}] = \int [\, W \,] \, d\vec{\rho} \implies [\mathcal{L}] = \int [\, W_{ij}, T_{kl}] \, d\vec{\rho} \, d\, m$. If $\oint [\, W_{ij}, T_{kl}] \, d\vec{\rho} \, d\, m = [\mathcal{L}] \in P_N$ $\implies \oint [\, W_{ij}, T_{kl}] \subseteq H_M$. If we breakup the resulting derivation by using the method of computational control such that H_M remains invariant; then the following LF is stipulated through internal control: $[T] \longrightarrow H_M$ and $[W] \longrightarrow H_M$ such that $E_{H_M} \geq 16$ TeV as stated in the GRS [The Grand Unification Scheme] Energy Scale, which is now denoted Ω, such that $[T] \cap [W] \longrightarrow H_M$. [5] Resulting in the following statistical approximation: $\Omega \gtrsim E_{T \cap W} \gtrsim 16$ TeV. We produce systemic analysis through ratio in relation to initial starting point all the way to the estimated Kardeshev Scale parameter within the giving time-frame in classical approximation.

References

[1] [2] Sanchez-Rey, Miguel A. Metaspace. Vixra.org: 2016.

[3] Sanchez-Rey, Miguel A. The Physicalist Program. Vixra.org: 2014.

[4] Sanchez-Rey, Miguel A. The Logical Structure of Space-Time. Vixra.org: 2011.

[5] Sanchez-Rey, Miguel A. Energy Scale of the Grand Unification Scheme. Vixra.org: 2016.

Lagrangian Vertex Operator for Electrostatic Background Field in Ω

Miguel A. Sanchez-Rey

Abstract

It's expected that electrostatic background signatures surmounts at the supersymmetric [SUSY] energy scale. These electrostatic background signatures is intrinsic to metamorphic space. In order to derive a Lagrangian vertex operator one must treat SUSY as a metamorphic phenomenon.

September 18th, 2016

SUSY and Metaspace

First one demonstrates the action integral for SUSY at $\mathcal{N} = 4$:[1]

$$I_{\text{SUSY}} = -\frac{1}{2}\int_M d^5 \, e^{i(wt-k\vec{x})} \sqrt{-g} \, [R - \frac{1}{l^2}] \; + \int_{\partial M} d^4 \, e^{i(wt-k\vec{x})} \, \hat{K} \sqrt{-h} \; - \int d^4 \, e^{i(wt-k\vec{x})} \sqrt{-h} \; \frac{1}{l^2}$$

One shows that the supersymmetric solution as equivalent to the Polyakov action integral linear sigma model for curve background fields. Such that:

$$I_{\text{SUSY}} = I_{\text{Polyakov}}$$

Then one relates the difference:

$$I_{\text{SUSY}} - I_{\text{Polyakov}} = \varnothing \implies \text{INTERFACE}$$

$$[\, -\frac{1}{2}\int_M d^5 \, e^{i(wt-k\vec{x})} \sqrt{-g} \, [R - \frac{1}{l^2}] \; + \int_{\partial M} d^4 \, e^{i(wt-k\vec{x})} \, \hat{K} \sqrt{-h} \; - \int d^4 \, e^{i(wt-k\vec{x})} \sqrt{-h} \; \frac{1}{l^2}] - [\frac{1}{4\pi l^2} \int \partial^2 \xi \, [\sqrt{g} \; g^{\alpha\beta} \, G_\mu(x) \; + \; \epsilon^{\alpha\beta} \, B_\mu(x)]$$

$$\partial_\alpha X \, \partial_\beta X \; + \; \frac{1}{4\pi} \int d^2 \xi \sqrt{g} \; R^{(2)} \; \Phi(x)] = \varnothing$$

INTERFACE $= V < L, X^\mu > = \varnothing$ which is the vertex operator that yields the following Lagrangian for SUSY phenomenon in metaspace [2]:

$$\mathcal{L} = \oint V < L, X^\mu > = \varnothing$$

References

[1] Sanchez-Rey, Miguel A. The Logical Structure of Space-Time. Vixra.org: 2011

[2] Sanchez-Rey, Miguel A. Logical Form In Favor of Long Equations. Vixra.org: 2016.

Computational Factorization of Variants [of Stringy] in Metamorphic Space

Miguel A. Sanchez-Rey

Abstract

At a certain point computation will initiate catastrophe. One must then impose computational factorization of variants [of stringy] in metaspace.

September 18th, 1206

Variants [of stringy] have a number finite number of factors. One states the Factorization of PHPR as:

1. $p[O] = n$

2. $p[O] < p[p[n]]$

3. $p[p[n]] < p(p_1[n] \# p_2[n])$

Each variants [of stringy] is an element of perfect number. Then each variant [of stringy] has a finite number of factors that are elements of perfect number. Every instant a variant [of stringy] is factorized computational complexity is reduced. One states the Lagrangian for SUSY phenomenon in metaspace as $\mathcal{L} = \varnothing$ so that prime factor is of prime [1]. The Definition of the Grand Unification Scheme charge monopoles are of prime factor [2]. For conformal manifold of bosonic algebra prime factor is of α'. Even then giving the canonical form $[\, , \,]$ is of prime factor which is analogous to its path integral form between two points in relativistic quantum space-time. Any calculating procedure must be restricted to prime or it will end in catastrophe. Even then with the TrH Theorem number can be said of $1 + 2 + 1$ prime [3]. All polynomials [from the Jones and Alexander polynomial to HomFly] can be a factor of prime. 11-dimensional super-strings in conformal super-symmetric geometry is a factor of prime. Prime is reducible to itself or by prime. This implies that variants [of stringy] is reducible to themselves or by prime which states that all variants [of stringy] are reducible to each other therefore catastrophe is avoided.

References

[1] Sanchez-Rey, Miguel A. Lagrangian Vertex Operator for Electrostatic Background Field in Ω. Vixra.org: 2016.

[2] Sanchez-Rey, Miguel A. Physics in the Grand Unification Scheme. Vixra.org: 2016.

[3] Sanchez-Rey, Miguel A. TrHT in the Grand Unification Scheme. Vixra.org: 2015.

The Quantum Field Theory Approximation

Miguel A. Sanchez-Rey

Abstract

One states the quantum field theory [QFT] approximation [QFTA] for the Physicalist Program [PHPR]. An approximation that relates current PHPR knowledge to second quantization of relativistic point-particle interactions and QFT.

October 3rd, 2016

Relating QFT to current PHPR knowledge is a step that translates variants [of stringy] to second quantization and QFT. Second quantization is measurable science that explains many quantum point-particle interactions. QFT relates quantum mechanics and special relativity. QFT applies the path integral and canonical formulation that inevitably yields the Dirac equation, renormalization and regularization of anomalies in point particle interactions, Higg's mechanism, stochastic quantum mechanics, further understanding of the Electroweak, Nuclear Force and Higgs Force, the graviton, and inevitably the super-symmetry [SUSY] phenomenon.

The procedure is to relate the primary matrix to the Wilson operator of variants [of stringy], and all other variants, which are elements of perfect number. It is to relate metaspace to the primary axiom. In doing so one translates each variant [of stringy], of perfect number, to QFT. Doing so yields astronomical advances in the engineering and technological sciences. But to do so one must realistically complete the grand unification scheme [GRS] as stipulated or no avenue is possible to achieve such applications [it is too unstable] [1]. Also keep in mind by applying the prime factorization of PHPR eases the complexity of such translation and by using the Definition of PHPR one uses logical form to impose computational control and SUPREME to achieve order in metaspace:

$$\exists\, Z\,(J) \;\exists\, l_{[\text{parameter}]} \;\exists\, \Psi \;\exists\, L_{m,n} \;\exists\, \tilde{\tilde{L}}_{\text{min}} \exists\, D[p] \dashv \mathcal{L} = \varnothing$$

[2, 3, 4]

References

[1] Sanchez-Rey, Miguel A. The Physicalist Program. Createspace: 2015.

[2] Sanchez-Rey, Miguel A. The Foundations of Quantum Field and It's Particulates. Vixra.org: 2014.

[3] Sanchez-Rey, Miguel A. Computational Factorization of Variants [of Stringy] in Metamorphic Space. Vixra: 2016.

[4] Sanchez-Rey, Miguel A. Global Policy and the Scientific Age in the Modern Era. Vixra.org: 2016.

The Sociopathic Norm: A Pattern of Organized Criminal Activity in the Scientific Age

[Author: Miguel A. Sanchez-Rey]

It is only recently that a pattern of behavior has been more tantamount. A pattern of behavior in line with organized crime in the Scientific Age. This pattern of behavior can be said to be an ingrained pattern of behavior that is hard to catch but when spotted can be frighteningly surreal and deliberate.

Organized crime in the Scientific Age can only be said to be more evolved and elaborate. With crime elements wide-spread in the Tri-State Area and New England States in the United States, scattered in Western Europe, prevalent in Latin and Central-America, and isolated pockets existent in Eastern Europe and South-East Asia and the Polynesian countries including, Australia and New Zealand, in other words, they are a world-wide criminal element. Crime that went world-wide in 2002 and has since ravished the planet by causing havoc to many academic institutions and financial markets. Even then it can be said that this crime element has since evolved from organize crime to a sophisticated crime ring that involves many, supposedly, respectable foundations and firms collaborating in organize criminal activity to further their fame and fortunes.

But why is it so hard to catch this cult? It's only for two reasons: [1] They are so utterly famous that it is hard to believe that their mediocre and fraudulent in nature. [2] They are so benign and silly, with claims of altruism and benevolence, that it is impossible to even believe that they are capable of horrid acts of hatred and malice.

One way to perceive this criminal element, and its subtle patterns, is to see them as a more vivid form of a Peoples Temple Cult. A cult so brutal and violent that it ultimately leads their followers to mass suicide when confronted by the futility of their insignificance.

The Sociopathic Norm resort to brutal tactics to control each other and others they lure into the cult. Those brutal tactics amount to: [1] Rape [2] Brainwashing [3] Control [4] Social Isolation [5] Sex Trafficking [6] Warp and eventually homicide. Warp can be said to be a tactic in which one destroys a subjects' sense of dignity by imposing a trick in their cognition in order to control them.

The attributes of this cult are, as explained, very subtle: [1] Clueless clues. [2] Plagiarists [3] Racketeering [4] Sycophants and and cronies [4] Memorizers [5] Paperback writers [5] Genocidal maniacs.

Clueless clues is a behavioral tactic in which the cult leaves clues to those they attempt to induct or lure into the cult. Clues which are gathered base on a past profile of the subject [or victim]. When the victim confronts the object that imparts those clueless clues they impose frequent deniability until admission of those clues leads to a violent response. It is a sociopathic trait that eventually results in a violent confrontation and the potentially unfortunate end to the victim.

But what happens when this cult gets caught? It's very easy to say that they can hide their pattern of behavior but very hard to see how? When the Sociopathic Norm is expose usually they are seen as a rabid threat but for strategic purposes they may hide their fraudulent nature by posing as loving and caring public figures but if only so closely do you realize their clownish nature. It is then that most may see harmony but in truth it is an insane planet that is rather strange and chaotic.

www.ingramcontent.com/pod-product-compliance
Lightning Source LLC
Chambersburg PA
CBHW081311180526
45170CB00007B/2664